T0072936

Save
Our
Freedom

Bijan Moini

Save
Our
Freedom

A wake-up call in digital times

Translation by Oliver Latsch

Arctis

W1-Media Inc.
Arctis US
Stamford, CT, USA

Visit our website at www.arctis-books.com

1 3 5 7 9 · 10 8 6 4 2

Library of Congress Control Number: 2020943692

ISBN 978-1-64690-008-4
eBook ISBN 978-1-64690-608-6
Translation by Oliver Latsch
Cover design based on the Design by
Hauptmann & Kompanie Werbeagentur, Zurich

Printed in Germany, European Union

For Lio.
His future.
His freedom.

"Freedom was an ideology that was common into the 21st century, according to which every human being was supposed to be free to choose between different options."

Page "Freedom"
in: Googlepedia, The Encyclopedia.
Processing status: 21.02.2120, 23:55 UTC

Contents

Introduction

In the digital age, it's easy to get swept up in online shopping, Facebook, Netflix, Wikipedia, online gaming, Twitter, and more. We feel free, as though we can arrange everything exactly as we like.

Or that's what we're led to believe. We're relentlessly driven by technology, breathlessly checking our pulse on our smartwatches, plowing through hotel reviews, snapping up our smartphones at every vibration, losing ourselves in a maelstrom of YouTube videos.

There is method to this madness and this method comes with a snag. And this snag will drag away our liberty.

Unless we act now, our lives will soon be determined by algorithms, rather than our own choices. But technology itself is not the problem: It's corporations that reduce us to mere metrics and manipulate us. It's the right-wing populists who lie to us and intimidate us. It's the State that monitors us and presumes us guilty until proven otherwise. They all put us under their yoke and, in return, satisfy our thirst for fun, for resentment, and for security. It's a very soft yoke—so soft, in fact, that we barely feel it—and all the while, it robs us of our agency.

But we can still stop the high priests of the algorithm from plunging humanity into a new Middle Ages, where the individual is once more subordinated—not to religion this time, but to superior machines.

To prevent this, we must (1) understand how our heedless pursuit of convenience leads us to sacrifice our freedom and dignity. We must (2) remember how precious that freedom is. And we must (3) find ways to salvage our freedom without necessarily relinquishing the benefits of digitization.

Yes, it *is* possible. We can build a digital future in which we live free, self-determined, and dignified lives. And we already have the tools we need, both politically and in our individual actions.

1.

Our Liberties are Threatened More than Ever Before!

For most of human history, the dominating forces impacting society were hunger, disease, war, and natural disasters. Gods and princes determined a community's identity. But a few centuries ago, people decided to take their fate in their own hands. The American Revolution, followed by a century-and-a-half of revolutions from France to Latin America to Russia broke the fetters of religion and monarchical rule, replacing them with the tenets of the Enlightenment and reason and the rule of the people. Above all, however, these shifts in authority placed the individual at the center of our thinking. Henceforth we—not any higher powers—were in charge of our lives.

The bourgeoisie and a scientific curiosity freed from ecclesiastical dogmas brought us three more industrial revolutions: mechanization, mass production, and information technology. Each of these developments caused major upheavals and, ultimately, rearranged the balance of power. Intellect displaced brute power. Machines displaced human labor. Industrialists displaced the nobility. Intellectual property displaced material property. Now, with digitization, we are already in the throes of the fourth industrial revolution.

Everything that you can now imagine is being converted into digital formats and networked. At home, in the car, in the office, in science, culture, and the media, digitization promises us a healthier life, safer and more comfortable transportation, more security, and, provided we behave ourselves, perhaps even an unconditional basic income.

But in solving old problems, our technology all too often creates new ones. Its greatest victim so far is the climate—and its next victim could be our freedom. We are told that to make our dreams come true, we too must be digitized. The databases of states and corporations create our digital alter egos, and we do not have the right to decide about our inclusion in them.

At the moment, no indicator lights up when our adrenaline levels are too high, nor does a signal warn our surroundings if we are in a bad mood. Even so, our many interactions within the digital world are being recorded, stored, analyzed, and used.

This makes us vulnerable. An entire industry thrives on predicting our behavior and seducing us. Everyone knows its most powerful players: Google, Facebook, Amazon. They've learned to extract gold from colossal data repositories. Artificial intelligence (A.I.) is their finest sieve. A.I. helps them collect even more data by capturing our attention through recommendations for more videos and news. Third parties increasingly know how to exploit this targeting for themselves. The most shameless of these is a resurgent right-wing populism, which uses Facebook, Twitter, etc. better than anyone to promote its rhetoric. Instead of preventing this, the State is, itself, busy measuring and prejudging us in the name of security. Everybody is recorded and categorized, increasingly often by algorithms.

Current law does not sufficiently confront these dangers to our freedom in our daily actions and thoughts. That is why, despite all the wonder at the marvels of new technologies, we must fear for our freedom as capitalism, right-wing populism, and the security state squeeze it in a powerful digital vise. Only the most flexible, transparent people will do well under this treatment.

Fundamental Rights Safeguard Our Freedom, but Not from Manipulation

For a long time, freedom was only for the elites, tribal leaders, priests and princes, plantation owners, and industrialists. It took a long time to abolish slavery and serfdom and introduce the rule of law and fundamental rights, and this transformation has still not been achieved everywhere. Nevertheless, more people are free today than ever before—not free to do whatever they want, but free from being subject to the will of others. Thanks to this individual freedom, we can openly articulate and assert our interests. We can be who we are, say what we think, believe and love what and whoever we want. We can see, read, and listen to what we like.

This individualized freedom is protected by our civil rights, which are the first and best line of defense against the State. The State must not take our lives or injure us. It must not restrict our religious practices or forbid us to speak. It must not touch our property or restrict our choices of profession—unless there is a good reason for doing so, and the infringement on our rights is proportional to the advantages to society as a whole. Moreover, the State is obliged to protect us from any violation of our fundamental rights by private individuals and companies. That's why the police are required to intervene if someone attacks us. And that's why the State cannot leave the operation of dangerous technology, such as nuclear power plants, to private companies without formulating specific rules for operating them safely.

These fundamental rights are enshrined in national constitutions, such as the Bill of Rights, the first ten amendments to the U.S. Constitution, or the German Basic Law, and many international treaties. Thanks to the right to adequate legal protection and representation, we can sue in national courts up to the German Federal Constitutional Court, the U.S. Supreme Court, or the European Court of Human

Rights in Strasbourg and the Court of Justice of the European Union in Luxembourg.

All this is only meaningful because the wielders of physical state power—the public administration, the police, the military—respect our rights and the judgments of the courts.

The prerequisite for exercising our rights is that we are free in another sense. Respect for civil liberties is a necessary, but not a sufficient condition for true freedom. According to the philosopher Hannah Arendt, we must also have the freedom to be free, which requires freedom from want.

In the digital age, being free is even more significant. It means being able to form one's own will, free from manipulation.

This dimension of freedom is under sustained attack from various quarters, and mostly goes unnoticed by us.

And as a consequence, our civil liberties themselves are crumbling.

Attacked by Artificial Intelligence

Information technology (I.T.) has developed at a breakneck pace. In just a few decades, it's come a long way, from the decoding of encrypted messages in the Second World War to simultaneous digital translation, and it has done so at an ever-increasing pace. The storage capacities have become significantly greater, the processors faster, the software codes more complex. Moore's Law states that digital processing power doubles about every two years, and this still holds true.

I.T. has produced a great deal that we wouldn't want to do without. Because of our smartphones, we always have a telephone, navigator, camera, bank teller, mailbox, games, and the whole internet in our pockets. With a few swipes and clicks, we can watch our favorite movies, shop, or order dinner. We can reach our loved ones anytime and anywhere, participate in the lives of strangers, or network with like-minded people through social media. Without advances in I.T., medicine would not have developed as quickly as it has, nor would engineering and art have advanced as they have. Law firms would still have to engage in extensive manual labor to search physical libraries for relevant decisions, companies would have to keep actual books for their bookkeeping, and bands would have to spend a great deal of money recording their first songs in studios. Flying would be less safe, and cooking dinner at home a lot more expensive.

All this is nothing compared to what artificial intelligence already does on a daily basis and what it will evolve to do. With everything digitized, in the future there may come a time where there is no problem we don't first attempt to solve using A.I. But how does artificial intelligence actually work, and what can it already do?

A.I. Can Only Detect Patterns—
A.I. Can Detect Patterns!

A.I. has only just learned to walk, and already it's leading to revolutionary strides around every corner. Self-driving cars are revolutionizing the management of traffic patterns. Smartwatches are revolutionizing the quality of health care that can be provided to patients. Data analysis and facial-recognition software are revolutionizing policing. YouTube and Netflix are revolutionizing the consumption of entertainment, Amazon is revolutionizing how people shop, Google is revolutionizing how they seek out information, while Facebook is revolutionizing how they communicate.

A.I. plays an increasingly consequential role in all areas of society, and yet it's not even particularly smart: it can't set its own goals; it can only perform one task at a time; and it requires human guidance for almost everything it does. Still, given sufficient data and computing power, it is vastly superior to humans in recognizing patterns. This may seem trivial, but it has unforeseeable consequences. A.I.-supported software can recognize animals, objects, and handwriting on photos. It can detect interest and engagement in click patterns and eye movements. And time after time, it beats the best in the world at chess, Go, and poker.

All this is possible because today's computers can process enormous amounts of data quickly and learn from it. In the past, humans had to define the rules by which computers operated. Today, thanks to A.I., machines learn the rules themselves—much as humans do—only much faster and more precisely. To better demonstrate this, think of a father who wants to teach his child what a cat is. He doesn't say *A cat is a living being on four legs with fur, a tail, and long whiskers on its muzzle that makes a* meow *sound.* Instead, he points to a cat and says, *That's a cat*, repeating that definition as many times as necessary. When the child points to a dog in the park or a cougar in the zoo and says *cat*, the father corrects them. New connections form in the child's brain until they understand what makes a cat a cat. From then on, whenever the child sees a cat, specific neurons activate that signal correctly, a *cat*.

Machine learning works in similar ways. Humans mark cats in countless photos and videos and use them to train the software, which uses a neural network modeled on the human brain. Almost everyone has helped train an A.I. at some point. For example, by solving so-called captchas, we prove that we are human beings by marking traffic lights or vehicles or recognizing a house number. The larger the data set used for training, the more reliably the machine will later identify the object in an unfamiliar photo. This is why large amounts of data (Big Data) are so valuable and why that technology is only now spreading quickly. This is what A.I. is today—nothing more, but also nothing less. And companies and governments

SAVE OUR FREEDOM 21

rely on it more and more. After all, the technology promises to perform any task according to a set of rules quickly, reliably, impartially, and without ever taking a break.

And there are quite a few rules.

The Promise of A.I.: A Carefree Life

Thanks to A.I., smartphone assistants understand our spoken commands, programs translate texts simultaneously, and Legal-Tech software finds the crucial clauses in complex contracts. A.I. determines which online advertising we see, when we see it, and in what form. It determines the price we pay for a flight, depending on the date, time of booking, and the device used. It teaches robots how to walk—and even smell—and it can already read minds.

The police use A.I. to predict apartment break-ins: until the summer of 2018, the police tested intelligent video cameras at Berlin's Südkreuz train station, designed to recognize the faces of registered "potential threats" in real time. Law-enforcement agencies are also using algorithms to analyze all international air passengers' data to identify people who might commit a crime in the future.

A.I. helps hospitals detect certain forms of cancer on X-rays and scans, and to predict when patients' conditions will worsen due to acute kidney failure. It can already diagnose mononucleosis, meningitis, and some other rare diseases. In some instances, A.I. seems able to determine a probable date of death using X-rays. A.I. will also make it easier and more efficient to analyze DNA, which will help develop individualized therapies for patients.

A.I. can help improve our quality of life in other areas, as well. For instance, self-driving cars provide additional safety measures that decrease the likelihood of injury or death when compared to the outcomes produced through human error. The A.I. utilized by these vehicles enables cameras to detect traffic signs, people, and obstacles. Fully autonomous traffic management should prevent accidents, mitigate the impact on the environment, save us from traffic jams, and, best of all, for most drivers, it's a cheaper option.

A.I. is already involved in finding us a life partner. Dating apps use A.I. to suggest people with whom we should be particularly compatible. To do this, they analyze the selection of other users with the same taste or bring together people with similar attractiveness values. Some take it even further: in 2018, a lonely Japanese man married a synthetic pop icon, Miku Hatsune, the virtual embodiment of an artificial singing voice. Thanks to A.I., robots have been used with some success in geriatric care as conversational companions, as well.

Some of these examples suggest the far-reaching and exciting potential of A.I. to even make work superfluous. If A.I. can find criminals, detect diseases, drive cars, provide partnership, care for the elderly, and so much more, why shouldn't we sit back, let machines work for us, and enjoy our limited time on Earth? Are we on our way to paradise?

No, we are not.

The Problem of A.I.: It Threatens Our Liberty

Everyone has secrets—things that have happened to us, actions we regret, information or experiences we're ashamed of. No one wants the world to know everything about them. But in a digital era, that becomes increasingly difficult. A.I. creates exact personality profiles from the information shared as we move in digital spaces—how much time we spend in them, where we spend that time, with whom we communicate. It can already read feelings from our facial expressions. And soon, it will be reading thoughts.

But do we lose something when we're all secretly monitored? When our interests, behavior, and feelings—even our thoughts—are stored and analyzed? When increasingly accurate conclusions can be drawn about our innermost selves? When our future behavior can be predicted? What we may lose is difficult to grasp, and that may be because it was never under threat before.

The highest German court, the Federal Constitutional Court, encountered that problem when it heard its first case about digitization. In 1983, it had to rule on the electronic processing of personal data raised as part of a nationwide census. The German Constitution's text did not and does not set any explicit limits on data processing, so the court formulated a new fundamental right that sounds unwieldy at first glance, but that hits the nail on the head: the right to informational self-determination. It guarantees the individual's right to decide, in principle, about the disclosure and use of their data. Self-determination is necessary for every human being's dignity, as guaranteed by Article 1 of the Basic Law (Grundgesetz), the German Constitution.

In the digital world, however, it is not we who make these decisions. Everything we have said, done, or even thought in the digital space is immortalized in our digital alter egos. We are split, having sovereignty over only the physical, while becoming intellectual subjects to digital potentates. We go from being a dignified subject to an available object.

The subject acts; the object is traded.

And trade is booming, because the digital world is in the process of displacing the real one. In that digital world, only our alter egos matter. We have to make do with the services and products offered, with search results and dating suggestions, news and songs. If our digital ego is considered dangerous, we'll have the police on our cases. Decisions are made by algorithms. An appeal is impossible—how can you argue against an algorithm if even its programmers no longer understand how it makes decisions?

Worse still, when we lose the possibility of deciding for ourselves, we will soon after lose the ability to do so, and with that loss, we lose confidence in ourselves.

In the past, few people followed the instructions of navigational devices. The first GPS devices led us into dead ends, did not recognize one-way streets, and could not reliably locate us. A printed map was always handy in the glove compartment or on the lap of our copilots. Eventually, however, GPS technology began to work so well that we preferred listening to them rather than to our copilots. Today, we don't even listen to our own navigational knowledge anymore. We used to trust God to show us the right way. Then we trusted our intellect. Now we trust Waze.

This trend also applies to tasks of much greater significance to society: police work, architectural design, medicine, and even software coding. For a long time, algorithms have been manipulating our perceptions through the internet, and soon they'll shape how we see the real world, through data goggles and virtual-reality filters.

Algorithms will register and track every step we take. And, along the way, they will enrich their creators' lives immeasurably, allowing them to profit from our very existence.

Attacked by the Capital

The history of humankind is the history of the struggle for self-determination. We fought for freedom from foreign domination and divine rules, from conventions and patriarchy. The most crucial battle will be against ourselves. Against our biology. Against our susceptibility to manipulation.

Self-determination is, strictly speaking, an illusion. We develop from a sperm and an egg cell consisting of little more than the codes that determine our development into a complete human being. Once in the world, we are shaped by influences we cannot control: family, friendships, school, conventions, laws, ideologies, experiences, stories, and culture. These influences shape the rules according to which our brains react to stimuli and trigger thoughts and actions. Neuroscience and philosophy have come to agree that there is no spirit, no soul—nothing that determines what we think and how we act that is independent of the physical-chemical processes in the brain. Instead, we *are* these very processes. And as long as they're free, so are we. But we can be influenced by those who understand the processes in our brains and can control the stimuli that act upon them. Therein lies the power of creating personality profiles and reading feelings and thoughts. And *that* is what digital capitalism exploits for its purposes.

Exploitation Through Manipulation

Technology is neutral. A knife can cut apples or kill a person. A plane can take us to a vacation or to go to war. A radio can spread music or hate speech. It is not the technology that's at issue; it's how it's used that makes it dangerous. The same goes for A.I.

Capitalism is not neutral. It promotes and demands the best and the worst in us. Its drug is competition. Its target is big money. Its drive is self-interest. It uses any technology that promises an advantage, and that includes A.I.

The distinction between the technology itself and its use by companies highlights the motives behind digital business models. Take Facebook, for example. An account on the social media platform costs nothing, nor does messaging. Facebook reminds us of anniversaries and special events; it presents us with personalized news and helps us keep in touch with old friends. The company takes care of us, almost lovingly. But at what price?

Price? you might ask.

In autumn 2019, the five most valuable corporations in the world were Microsoft, Amazon, Apple, Google's parent company Alphabet, and Facebook. In 2018, these companies achieved a combined annual profit of an incredible $139 billion. The smartest minds on the planet work for them, no one can escape their products, and the United States is proud of their achievements. It's a problem because pride doesn't mix well with skepticism, and more skepticism would have been appropriate over the past twenty years. Perhaps there would still be a few A.I. brains left to compete with these I.T. companies' lavishly equipped research departments. Perhaps some intrusive innovations would not have crept into people's everyday lives.

Facebook earned its 2018 $22 billion profits almost exclusively through advertising. This is only possible because Facebook promises its customers—by which I mean advertisers, not users—effective ad placement. Google promises the same thing. Amazon suggests products we did not explicitly search for, and purchases based on these recommendations account for more than a third of Amazon's sales. In short, these companies make money by promising to sell us something.

But we are not the product, as many claim—we are a resource. Facebook et al. achieve the high effectiveness of their product suggestions by analyzing and manipulating us. Algorithms learn who we are, what we like, and what we are interested in. They understand when we are most receptive to suggestions, and make us suitable offers at exactly the right moment. The more accurate their predictions are, the more money they can charge their customers for ads. But for this, they need more and more detailed personality profiles from us. They want to know, among other things, our incomes, our preferences, our current situations in life, and our weaknesses. In 2017,

Facebook monitored 6.4 million young people in Australia and New Zealand, and determined when they were stressed, overwhelmed, or anxious. The company used this "ability" to advertise its ad placement services.

One might argue that advertising has been around forever in the form of street vendors, newspaper ads, TV commercials, and more. Advertisers have also always sought to identify their target groups, preferably by choosing the right medium, the right message, and the right timing. But personalized advertising now has a new quality. It makes a big difference whether we belong to just one target group or whether a company can address us in precisely the right way, with precisely the right advertising, at precisely the moment when we are most receptive or weak. This form of advertising crosses a line because it sees and recognizes us, and exploits these insights in an invasive manner.

And this capability is not solely used for advertising.

The Danger of Concentrating Power in Corporate Giants

Product advertising shows what Facebook et al. are capable of, but an even greater danger arises from the use of A.I. for other purposes. For example, Facebook can influence people's state of mind. One experiment showed 300,000 users negative messages from their friends, and found that it made their moods worse. In another experiment, Facebook was able to increase voter turnout in the 2010 U.S. Congressional elections by two percent by introducing a simple feature that indicated to friends when they had voted. Others are using Facebook's potential even more aggressively. For example, Cambridge Analytica alleged that in the 2016 U.S. presidential election, undecided voters were drawn to Donald Trump's site en masse with tailored messages.

Despite its already precarious power, Facebook is thinking ahead. The company is building a global state: 2.9 billion people use Facebook platforms, including the social network itself, as well as WhatsApp and Instagram. The company has offered access to the internet to remote regions of the world—but only to sites that Facebook would control. It's announced its own electronic currency, Libra, which could turn the world's financial system on its head. Libra is supposed to enable users to pay across national borders, not only on Facebook but in clothing stores, for meals in restaurants, and other transactions, all without exchange rate fluctuations and fees. It's unclear what will happen with the data from these payments.

And all of these examples relate to only one company. Facebook is one example of the power that the owners of the big I.T. companies—each a monopolist in its field—exert over our lives. Google not only offers the most widely used internet search engine, but also Android, the most commonly used smart-phone operating system, and with their ownership of YouTube, the most popular video platform. The group is also investing in A.I.-supported healthcare and is developing a self-driving car. Amazon, in turn, is not only the largest online market, but also sells the most popular smart speaker, Echo, and operates by far the largest server farms in the world. These are gigantic warehouses full of computers, whose compu-ting power is used by many other companies, and no-body can avoid their influence in everyday online life.

In turn, these corporations influence at least half of the traditional economy. If a PR firm, a restaurant, or a toy manufacturer doesn't create opportunities to get a Facebook like or a Google review for its prod-uct, it loses.

Once the I.T. companies have harvested the avail-able data, they widen their net by recruiting new us-ers. And the bigger and bigger these companies grow, the easier this harvesting becomes. They swallow competitors, develop tempting apps, and buy data from third parties. And they keep us on their plat-forms longer and longer. Getting likes on Facebook and Instagram is addictive; on Netflix, the time be-tween one episode and the automatic start of the next is only five seconds; more and more information is directly available on Google's search results page, so you don't have to follow the links to other pages; YouTube's recommendation algorithm is responsi-ble for over 70 percent of the time we spend there.

And during this time, YouTube—as *The New York Times* has reported—systematically leads us into ever deeper abysses: to conspiracy theories, extremist propaganda, even to videos that sexualize children. All because the algorithm has noticed that this keeps certain people on the platform longer.

I.T. companies store our digital identities. In return, we are allowed to search the internet, scroll through Facebook timelines, and watch funny—or not-so-funny—videos. That's not much in return for us giving up our digital autonomy.

Other entities also earn money with our data. U.S. economist Shoshana Zuboff calls this new order "surveillance capitalism." This term, however, underplays the problem, because it's not only about surveillance. It's about manipulation.

Digitization hijacked by capitalism has become an instrument of power that facilitates our exploitation. And capitalists aren't the only ones profiting from it.

Attacked by the Far Right

Right-wing populism has been on the rise for the past decade. It's dismantling liberty in the heart of the European Union, Poland, Hungary, the Czech Republic, Austria, and, in the form of Brexit, in Great Britain. Its poison has had the most dramatic effect in the United States. The stereotypical bogeymen and rhetoric used in these countries are strongly reminiscent of the period of fascism in the first half of the twentieth century. Right-wing governments replace judges, penalize minorities, change the right to vote, and—if their legislation violates the nations' constitutions—lead to the alterations of those very documents. Even where right-wing populism is not in power, its aggressive language shapes discourse and policy. It also claims human lives, not only in the U.S., as in the killing of eleven Jews in a Pittsburgh synagogue by a right-wing extremist in October 2018, but also in Germany, where nine people with immigrant background were killed in Hanau on February 2020.

Digitization is crucial to the rise and empowerment of right-wing populism, which uses its tools against the one thing that burns like the sun in its eyes: the truth.

Misuse of Digital Communication Tools

According to one widely used definition, populists succeed by juxtaposing a corrupt elite with a virtuous people and declaring the latter to be the only legitimate source of political power. Their current most prominent representative is President Donald J. Trump of the United States. He spreads racist, misogynist, and nationalist messages to thousands of fans at "rallies," which are also broadcast on national television, and that are, more importantly, shared widely on the internet. No active politician in the world has more followers on Twitter. Additionally, hundreds of thousands of retweets and reports of his worst outbursts are in the traditional media and online. According to a survey, Trump reaches more than three-quarters of the U.S. population. Right-wing populists in Austria, France, Germany, Italy, the Netherlands, Poland, Sweden, the U.K., Brazil, India, and Myanmar use similar messages and channels.

No other political movement uses social media more successfully than the right-wing populists. In Germany, for example, according to one study, at the end of 2018 and moving into 2019, 32 percent of all Facebook and Twitter posts mentioned the right-wing extremist party Alternative for Germany (AfD), which only polls at 13 percent. At the same time, the Green Party, which polls at 19 percent, was only mentioned in 7 percent of posts. A similar picture emerged in Spain regarding the right-wing extrem-ist micro-party, Vox, and in Italy for Matteo Salvini's right-wing radical party, the Lega.

These developments point to the fact that right-wing populism has particularly benefited from how digitization is undermining the guardian function of traditional media. Its formerly ostracized messaging now reaches the electorate directly through posts on Facebook and Twitter. Its disproportionately active supporters reinforce the messages on opinion boards on the internet and via the comment functions of var-ious online media. As a result, international alliances of nationalists emerge, which mutually reinforce and radicalize each other. *The New York Times* has reported on how right-wing Swedish Democrats are supported financially and ideologically by right-wing conservative forces in the U.S., as well as by Russian propaganda channels.

Right-wing populists also shamelessly exploit the mechanisms of outrage in traditional and social medias to give their messages greater reach. News spreads faster when it arouses emotions. According to one study, each emotionally charged word increases the chance of a retweet by 20 percent. The key to right-wing populism is words or groups of words to which supporters and opponents alike respond: *illegal immigrants*, *Islamization*, *terrorism*, *fake news*. Or it coins terms by describing groups of refugees as a *wave* or *flood*, thus evoking images of uncontrollable forces of nature. This process is called *framing*. It shapes and influences the entire public discourse, especially through automatically controlled user accounts—so-called "bots," which disproportionately spread right-wing populist messages.

There are drastic consequences to this rise. Social media posts of the German AfD are demonstrably linked to violent acts aimed at refugees. Before the referendum, Brexit proponents placed advertisements in social media that stirred up unfounded fears of immigration or falsely blamed the European Union for the economic plight of disadvantaged groups in Britain. Russian troll factories made every effort to influence the 2016 U.S. presidential election in favor of Donald Trump. And Myanmar's military used Facebook to launch rapidly spreading racist hate campaigns against members of the Rohingya, a Muslim minority, which led to genocide. The best-documented example is the decisive contribution of YouTube's recommendation algorithms to the ascent of the right-wing extremist Jair Bolsonaro to Brazil's presidency. *The New York Times*, referencing various studies, traced how YouTube shifted the entire political discourse in Brazil in favor of right-wing conspiracy theories. It systematically led users to their disturbing theses. Even the election winners themselves openly admit that they owe their surprising success to the video platform.

Anyone who votes, polls, or even kills under this kind of influence is not exercising their own free will.

No Freedom Without Truth

Only those who know the truth can decide freely to opt for a risky medical procedure, against a job, or for a political party. So, among other things, when Donald Trump lies, it's an attack on freedom. *The Washington Post* counted no less than 16,241 false or misleading allegations made by Trump in the period between Trump's inauguration and January 2020. He invents "invasions" by "illegal'" refugees; he denies human responsibility for climate change; he boasts of never-achieved successes, including the lie that he got rich on his own. Before his election, he promised, in true populist fashion, to drain the "swamp" of lobbyists and corrupt politicians in Washington, D.C. for the benefit of the ordinary people. No U.S. president before him has lost as many cabinet members due to ethics rules violations. Anyone who exposes his lies is denigrated as a disseminator of "fake news," in a particularly ugly inversion of reality.

When politicians of the extreme right-wing FPÖ in Austria spread the lie that the Jewish U.S. billionaire and philanthropist George Soros was going to organize mass immigration into the E.U., they manipulated voters in the same way. The same applies to the German AfD. Ninety-five percent of its press releases between January and October 2018 were about criminal acts allegedly committed by foreigners. The former Italian Secretary of the Interior Matteo Salvini declared maritime rescue operators in the Mediterranean to be "accomplices of smuggling gangs," and refugees to be criminals. And they all manipulate voters when they promise to return their countries to their former strength.

Demonizing the press, creating bogeymen, fomenting fear, and glorifying the past are all tried-and-tested tools of right-wing propaganda, to which digitization lends new clout. Like capitalism, right-wing populism exploits the fact that we are capable of being manipulated. In doing so, it pokes a finger into a chronic wound of democracy: it captures people's anger at an unjust world, their fear of the

unknown, their desire for recognition, and their belief in a conspiracy against their peers. Thus aroused,
those converted to the touted ideology no longer
make decisions freely, nor in their own interests. For
instance, the many people who fell for the battle cry
"Take back control!" during the Brexit campaign
now realize that when trade slows down and jobs
disappear, when you no longer count for anything in
the world because your country is isolated, you may
luxuriate in self-control, but not in security and prosperity. Or look at those sections of the U.S. working
class who voted for Donald Trump. His tax policies
favor the rich and his trade policies hurt the poor, his
cabinet is full of millionaires and he is filthy rich (or
at least he claims to be).

Those who do not fall for the lies of right-wing
populists are crushed, especially where these populists hold power. No one in the European Union
dominates his country's public discourse more than
Viktor Orbán in Hungary with his Fidesz Party. In
a public speech, he announced the end of liberal democracy and praised the illiberal systems in Turkey,
China, Singapore, and Russia. He restructured the
state media until they distributed only his propaganda. Oligarchs close to him acquired all the essential
private media and brought them into line with the
government. Nongovernmental organizations found
the work so difficult that some of them had to close
or move away. One man, one party, one truth.

But aren't the Western democracies prepared for

all this? Doesn't our state protect freedom? Doesn't it stifle the calls of capital and populism into digital space?

Not at all.

Attacked by the State

Mother Capital is the driving force behind turning humans into metrics, but Father State has long since joined her. He collects vast amounts of data, searches for patterns, and automates his decisions. Belgium is trying to identify and summon unmotivated unemployed people electronically. Denmark wants to use data analysis to identify neglected children. The most significant impact of digitization and the use of algorithms, however, is on security policy. China will soon have completed the digital recording and control of its population. A mixture of technological ambition and the Communist Party's unquenchable thirst for power is drying up all the tributaries to the benefit of the great river of system conformity.

All this could also be coming for us, too. If we extrapolate technology and its use today, then freedom will soon be nothing but a historic encyclopedia entry.

Security Above All Else

Shortly after the planes exploded in the World Trade Center on September 11, 2001, the world saw an explosion of security legislation. Starting in the U.S., new criminal offenses and higher sentences were introduced, and expanded powers for secret surveillance and the police were established—with a little more latitude each year. And increasingly, these powers were, and are, about data.

The BND, the German foreign intelligence agency, with full legal authorization, now scans any communication outside of Germany that it can get hold of. (They were doing it secretly until Edward Snowden's revelations came out.) Only a ruling by the European Court of Justice prevented the storage of national telecommunications data without cause. The German Federal Criminal Police (BKA)—the German equivalent of the FBI—is currently setting up a gigantic database that contains many individuals who have never committed a crime, but are, or have been, suspected of one. And this database is growing every day. The BKA also stores and processes extensive data on all passengers on international flights, and is planning to include bus, train, and maritime travel data. The state police authorities can use spy software to monitor potential threats as they chat, message, or search computers and smartphones. The principle of separation between police and intelligence services—a consequence of the experiences of the Gestapo in Nazi Germany and the

Stasi in East Germany—has effectively been abandoned. In the fight against terror, authorities are allowed to monitor us at an even earlier stage: Whereas previously an apparent threat was a prerequisite for intervention, in Bavaria, an "imminent" threat is now sufficient. This means that the police have swapped their small net, which they were required to use only to clean the aquarium, for a large one, which will catch more algae, but also fish.

Liberty has become an inconvenience to the state. Yet the purpose of security should be only to ensure liberty. Algorithms help to mark potential threats. They search for "suspects" in passenger data records based solely on their flight behavior. The German state of Hesse is utilizing software from the U.S. company Palantir, which the police can use to search various public sources and the entire internet for a person, and then display their digital network. In the U.S., this software, known as Gotham, has access to so much data that, based on a name alone, it can create a profile of a person's movements, provide e-mail addresses, telephone numbers, addresses, bank details, a social security number, business relationships, family members' information, and body measurements. And it's not just U.S. law enforcement using Gotham; the Secret Service and the Pentagon also utilize the program for analysis.

Underpinning the employment of these purported "miracle cures" is a desire for absolute security. This desire is often stirred up by politics itself, fueled by extreme right-wing populist demands. The State wants to stop people before they harm us. Sounds good, until suddenly a particular flight pattern becomes suspect, or a visit to a mosque sends up a red flag, or talking to the wrong man at a protest is deemed unusual. Everyone is under general suspicion, and the pressure to blend in continues to grow, and so we censor ourselves, refraining from making snarky jokes via e-mail or having a Facebook friendship with an activist, so as not to be interrogated at a police checkpoint or rejected at the U.S. border. This chilling effect is already happening. For example, since Edward Snowden's revelations, people have avoided Wikipedia entries that they consider suspicious. Furthermore, a significant proportion of the press leaves topics untouched that could make them the target of state surveillance.

The digital security state threatens our freedom even more than digital capitalism does. And this doesn't happen only when an extreme right-wing politician runs the responsible department or ministry. Even moderate politicians deliberately write complex laws, as the conservative German Interior Minister Horst Seehofer admitted only half-ironically in June 2019. They play with our fears, and then demand new and more far-reaching powers. In doing so, they systematically push the limits of their countries' constitutions, often moving beyond their granted authorities. A speed limit on motorways is a hot-button issue in German politics, but it would save far more lives than any anti-terrorism law. The same, arguably, applies to American gun laws. It is hardly possible for a civil society to counter this. And if complaints do arise, it is not easy for any Constitutional or Supreme Court to reject every single disproportionate restriction of freedom. The court's legitimacy partly depends on not placing too many limitations on politics, and not doing so too frequently.

Soon, anyone suspected of even thinking about a crime could have their reputation tainted and lose their freedom, and this could happen automatically. Algorithms are already involved in courtroom decisions. In the U.S., A.I.-supported programs help courts make decisions about granting bail, suspending sentences, and the application of other penalties. The individual—the specific case—falls by the wayside. At the same time, an effective defense becomes almost impossible because the algorithms' details are considered a trade secret worthy of protection.

If only our genetic makeup was granted the same kinds of protections! In the U.S., millions of people send tissue samples to companies like 23andMe or FamilyTreeDNA to search for distant relatives. These seemingly harmless services store, analyze, and exploit what is perhaps the most intimate thing we possess: the code that makes us. The FBI now compares DNA traces from crime scenes with these genetic databases. It rarely finds the person who left the DNA trace (who may not even have committed the crime), but it does often find relatives.

U.S. authorities are already using A.I. for automated facial recognition, even for minors, despite the risk of discrimination or false suspicion. For years, without the public's knowledge, they have compared pictures of wanted persons with over a hundred million driver's license photos to identify and deport undocumented migrants. The company Clear-View provides hundreds of law-enforcement agencies with an enormous photo database, which it continuously compiles from Facebook and YouTube. Amazon offers images captured by its Ring front-door camera to local authorities. One manufacturer of stun guns has resisted the integration of facial recognition into the police body cams it also produces. But it is probably only a matter of time before a stun gun can fire autonomously when a body cam detects an alleged attacker.

The U.S. military has long been researching the use of A.I. in selecting military targets, and it could soon control attacks by swarms of autonomous combat drones or mini-ships. It's worth thinking about what a war would look like if the other side also used A.I., since that other side would, more likely than not, be China.

Looking East: The Controlled Individual is Already Here

The People's Republic of China has achieved incredible advances in the last forty years. Since Deng Xiaoping opened the country in 1978, it has freed more than 500 million people from abject poverty, becoming first the world's workbench, and then one of its largest think tanks. China invests a lot in A.I., which makes sense since, due to its sheer size, the country sits on a gigantic treasure trove of data.

China's I.T. companies collect far more personal data than their U.S. counterparts. WeChat by Tencent started as a chat program along the lines of WhatsApp. Now, consumers use it to reserve tables in restaurants, arrange doctor's appointments, order food, look for jobs, apply for visas, and play computer games. More than one billion Chinese people use the app, and almost as many use the services of Tencent's competitor, Alibaba—a company that's eBay and Amazon wrapped in one, which also grants loans, organizes concerts, streams music, runs an online pharmacy, and more. However, the most influential inventions of these two companies are WeChat Pay and Alipay, which have made the smartphone a means of payment. Because of them, cash has become rare in China. Even donations to street artists or panhandlers are made via smartphones. In 2018, the volume of mobile payments issued via the two services exceeded 30 trillion dollars.

Through their cross-sectional offerings, Tencent and Alibaba know practically every detail about their users. In contrast to the West, all this data flows freely into the hands of the Chinese state. And this state is not overseen by a government elected by the people, but by the Communist Party. And it is the Communist Party that is pushing the digitization of the country vigorously forward.

It all begins at school. In a model project, school-children's meal orders were stored and evaluated, followed up with nutritional tips. Facial-recognition cameras registered late arrivals, attentiveness in class, and book borrowing. China is installing hundreds of millions of intelligent video cameras, some of which can recognize people not only by identifying their faces, but also by distinguishing their gait or cloth-ing. The police are already using data goggles to scan passers-by in the hopes of identifying those suspected of crimes. This is why in the summer of 2019, dem-onstrators in Hong Kong did everything possible to conceal their faces. Very soon, no one will be able to move around the country undetected.

However, the infrastructure and the data supplied by the large I.T. conglomerates are merely build-ing blocks of an even more monstrous project: the social credit system. Dozens of model projects are currently competing with one another, and one—or perhaps several—will soon cover every person and every company in China. One of the most prominent tests is running in Rongcheng. In the city and its surroundings, every citizen received 1000 points at the launch of the initiative. Accompanying older people to the market or sweeping the street earns you points. Throw trash in the street, cross against a red light, attend unregistered churches, or cause problems in the neighborhood, and you lose them. Criticizing the government costs the most points, placing investigative journalists and civil rights organizations in a particularly precarious position.

If an individual's social credit drops below a specific limit, they, for example, no longer qualify for a mortgage. In other point systems, those affected may no longer be allowed to fly or use long-distance trains, a consequence that has already impacted the lives of more than 13 million people as of spring 2019. The consequences include difficulties when traveling abroad, choosing a partner, and in the workplace. Because being acquainted with "dishonest" people may hurt your score, those affected may also lose their social contacts. Public denunciation by the authorities is widespread. A digital map shows people, in some instances, who have not paid their debts. Similar rules would apply to companies.

In contrast to credit rating systems used in the West, the social credit system is subject to sanctions with no material connection to the behavior being punished. In the U.S., anyone who has a poor credit rating will not be granted credit or may have to pay a higher interest rate. Bad reviews reduce book sales. But anyone with low social credit in China can lose on all levels: professionally, financially, privately. Control is perfect, and the Communist Party is the principal beneficiary in the long term.

The Uighurs, a Muslim minority in the northwestern province of Xinjiang, are already feeling the full force of the Chinese surveillance state. The state has stored DNA, blood types, fingerprints, and iris patterns of all Uighurs between the ages of twelve and sixty-five. Their movements are tracked by millions of smart cameras, not only in public but in some cases also at home. This isn't just happening in the provinces; intelligent video cameras perform automated racial profiling to recognize and track Uighurs using their facial features even in the wealthy cities of eastern China. The government allegedly also infected a large number of smartphones with malware that can read messages and passwords. According to estimates, one in ten of the 11 million Uighurs is, or has been, interned in a reeducation camp designed to exorcise their faith.

The treatment of this ethnic group is shocking, and it shows that total control is already possible. The people of Xinjiang have no room for resistance. They cannot hide or associate; they cannot even think freely.

Even ignoring such radical excesses, we cannot merely dismiss China. What happens there concerns us more than we might like. The country has become so rich and powerful that we cannot ignore its aspirations for the future, just as the world could not ignore the United States in past decades. Hardly any country dares to oppose China. Its market is too attractive, its reaction to criticism too relentless. The New Silk Road is the largest infrastructure project in history, building roads and ports on half the planet. In Africa, China is more present than any other country. It invests in America and Europe, and is the largest trade nation in the world. On the road to 5G technology, there is no avoiding the Huawei Group. TikTok is conquering our children's smartphones. And China's exports include A.I.-based surveillance technology. The country is even sending police officers abroad to Italy and Serbia.

Those who create values get to define them. And so, China's digital future could one day be ours: an ultramodern country without crime, but also without freedom.

But . . . would that be so tragic? Aren't prosperity and security more important than freedom? Aren't the 20 percent of Germans who, according to a recent report would like to see a surveillance state, perhaps right?

2.

We Must Value Freedom!

Some things are not appreciated until they're lost. When we're sick, we want to be well again. When we're bored, we want to be entertained. The same applies to freedom. Only when we are forced to do something for no reason, or when we observe such coercion happening to others, do we appreciate our freedom. We must always remember what we have and why it's important.

A dignified life is only possible when we do not fear that the State will punish our expression of opinions, arbitrarily deprive us of our property, or torture us. Self-determination in daily life is also incredibly important. In the first chapter, I spoke of its inadequate protections and impending loss. Self-determination brings happiness. That's why we must not be manipulated by I.T. companies, and must never allow the State to look inside our heads—not for comfort and entertainment, and not for safety.

Individual Freedom is
a Prerequisite to a Dignified Life

Modern autocrats don't wield power as clumsily as Hitler or Stalin did. They don't kill their opponents, rarely throw them in jail, and prefer to launder money instead of brains. Vladimir Putin has ruled in this manner for twenty years, and this is how Viktor Orbán transformed Hungary in a snap. It's also what U.S. President Trump is trying to do. That we, the people, are being disempowered by the modern right is no longer just a possible scenario.

We stand to lose a lot; civil liberties aren't simply a luxury accessory for affluent societies. They serve the highest purposes: life, liberty, and property. We see this in the first ten amendments to the U.S. Constitution, the Bill of Rights, which enshrined these very liberties in 1789. The same was tried in Germany in 1848 in the revolutionary constitution formulated in Frankfurt's St. Paul's Church, but it was never adopted. At the time, Germany was not a federal state, but a loose confederation of fiefdoms; it was not yet a democracy, but a union of feudal states and free cities. Arbitrary justice and unbridgeable differences in status dominated life. Over a half a century earlier in 1776, the United States had left those very conditions behind when it declared its independence from Britain.

There was, and still is, a constitutionally enshrined right to assemble and unite to protect one's social standing, property, and even life. The same holds true of the right to freely express one's opinion and to be informed by an uncensored press. Requiring arrests and search warrants, in turn, safeguards the freedom of the individual and the inviolability of the home. The freedom of religion protects the unhindered practice of one's faith. The independence of the judiciary enables the effective enforcement of the law. However, the most central right is the right to participate in the shaping of society through freely elected representation, for there is no better protection against tyranny than democratic participation.

Freedom and security have never been at odds. Freedom grants security. And the very listing of the few rights mentioned here shows that individual freedom is a prerequisite to a dignified life. Only those who are free from the arbitrary will of others can thrive and live up to their full potential. And those who live up to that potential are sovereign beings rather than the disposable assets of others.

And if all of this hasn't convinced you, let the many people who fought for freedom persuade you. The American Revolution of 1776, the French Revolution of 1789, the peaceful revolution in East Germany two hundred years later, and the Arab Spring of 2011 are only a few examples from the countless violent and peaceful, successful and unsuccessful revolutions and uprisings in human history. And all of them were about one or some form of freedom. Most were about the rejection of illegitimate rule.

Even governments that seem to offer their people everything are not safe from mass protests. The current democratic movement in Hong Kong proves this. The people living there enjoy many freedoms. The city is prosperous, but the wealth is unequally distributed. However, the people aren't demonstrating for economic justice. Rather, they're seeking the right to determine their own destinies. The government of mainland China refuses to allow free elections, although it had promised them when it took over the former British colony in 1997. This fact alone has driven millions into the streets, despite the menacing posturing of the Chinese military.

Even in the West, people may say, *I accept restrictions on others' rights as long as I'm safe.* But no amount of security can outweigh the suffering caused by oppression. The very essence of liberty is that it protects minorities from the tyranny of the majority. Anyone who thinks they're safe in the majority is deluding themselves. In some way or another, we are all in the minority and must trust in the universal application of civil liberties. The rich want to protect their wealth, the poor their livelihood. Christians defend their holidays, Jews a practice of religion without fear. Gun rights activists fight against restrictions, hooligans against stadium bans. Brad also benefits from Mohamed's freedom.

And what is right on a large scale applies on a small scale as well.

A Self-Determined Life is a Happy Life

Eric Schmidt, who ran Google for ten years, once said, "I think most people don't want Google to answer their questions. They want Google to tell them what to do."

We don't want that. The urge to take control of our own lives is something we were born with. Even toddlers are frustrated by boundaries. Teenagers curse their parents' rules. Adults don't want to be told which profession to pursue or which spouse to choose. We aren't ants or bees that happily follow their predetermined roles. It would go against our very nature.

Everyone has experienced the uplifting feeling that comes from making your own decisions—in your apartment, on vacation, after a separation, or after being fired. Or the joy of simply being oneself without having to meet the demands of others. When we do something of our own free will, we're more motivated and feel better. Happiness may also be found in daydreaming, trying something new, or simply doing nothing at all—in having time for all that, being unobserved, uncontrolled. Yet young people nowadays can't do, say, or write anything without having to worry about having to justify themselves to someone at some point.

That freedom creates happiness is not a mere anecdote: we have proof. According to various studies from various countries, economic and political liberties are positively correlated with people's sense of happiness. This is true worldwide, albeit to varying degrees. We know from experiments in retirement homes that it's primarily the freedom to make one's own life decisions that makes people happy. Therefore, freedom is one of the critical variables in the perception of the quality of life according to the World Happiness Report. Even the mere belief in free will strengthens our sense of happiness. According to a Chinese study published in 2017, this applies even in China. Consistent with the findings of various studies, our level of happiness moves along a U-shaped curve during our lives. We are happiest at the beginning of our adult life, the unhappiest usually between forty and fifty, and then our happiness starts to rise again. This phenomenon has a lot to do with our disappointment at not meeting our personal expectations and the fact that we are subject to the most constraints in the middle of our lives: financial obligations, pressure at work, and responsibility for children.

When digital capitalism takes decisions about movies, meals, or even relationships away from us, when right-wing populists distort society and truth with hate and lies, when the security state follows us wherever we go and suspects everyone, we become unhappy. And we'll become even unhappier over time.

Things can, of course, also take a turn for the better, but only as long as we are free.

Only Free Societies Evolve

Social change is like technological change: it's neither intrinsically good nor intrinsically bad. But without change, there cannot be progress for the better. We would still be enslaved and not free, and we would still be toiling in the fields instead of reading this book. We would still be discriminated against and accepting the status quo, rather than striving toward equality. This kind of progress required a certain amount of mobility, both in our circumstances and in our minds.

But every motion needs an impulse.

In a world determined by algorithms, every impulse is missing. Algorithms are structurally conservative; they learn from past data to shape the future. An algorithm that evaluates the probability of a U.S. criminal relapsing may not recognize that Black people are disproportionately charged and incarcerated more often, and that the data is therefore skewed to their detriment. A recruiting algorithm cannot acknowledge that men earn disproportionately more than their female counterparts. If nothing changes, racism and gender discrimination become entrenched. Algorithms don't go through an adolescent rebellion against conventions. They don't experience distress that drives them to revolt, nor do they suffer injustice that leads to calls for change. We may complain that some courts punish the same offense more severely than others. Still, in this unequal treatment lies a chance that, in time, a more effective response will prevail over the traditional reaction to a violation of the law. Equality for LGBTQIA+ people has more often been achieved by a changing interpretation of existing laws by courts than advances through legislation.

Our society needs freedom. The state must leave room for discussion, for deviant, borderline—even boundary-defying—behavior, because it is essential for science, culture, and politics to advance. If the equality of all human beings, if the right to participation, if the emphasis on individuality above the collective, if all those concepts had not been thought, discussed, formulated, and promoted, they might never have been able to rise beyond being treated as high treason, insult to the monarchy, and blasphemy. Without room for change, we would not be free now.

It is for all these reasons that we must act to prevent the loss of our freedom!

Only, how?

3.

We Must Save Freedom!

Technology creates demand. We need a car because cars exist. We need the internet because the internet exists. And we need the many services offered by the large I.T. corporations for the same reason: because everyone uses them. But what we don't need is their hunger for data. We don't need digitization at the expense of our freedom. We don't need to go from being humans and citizens to being reduced to data and consumers. If we had a freedom-saving alternative to Facebook, Amazon, and Google, we would use it. But alternatives are rare.

The German writer Erich Kästner once said about Nazi Germany, "One must not wait until the fight for freedom is called treason. One must not wait until the snowball has turned into an avalanche. One must crush the rolling snowball before the avalanche becomes unstoppable. Because the avalanche will not stop until it has buried it all."

In the first two chapters, we learned that we must overcome our indifference toward abstract, seemingly distant dangers. It was indifference that allowed the recognition of the aging of our societies to mature into a clear and present danger for our social safety nets. Indifference allowed the warming of our planet to develop into a catastrophe that is now almost impossible to manage. We must not let the same happen to our liberties!

It would be easy to preach renunciation now. But renouncing everything digital is not a solution. It is too simple, because it does not reach beyond the individual, and it is too hard, because it hurts us the most. We should not have to give up social media, online shopping, or smartphones. Nobody would demand cyclists walk if they were afraid of trucks. The solution lies in rules that tame the strong and protect the weak.

It is no coincidence, but rather a P.R. triumph by its opponents, that regulation has become a dirty word. However, regulation can help us get a handle on the many forces that seek to manipulate us. Balancing our weaknesses with rules and regulations is nothing new. Without rules, we would kill with impunity or raid other countries for greed. Without regulations, nuclear power stations would blow up and planes would drop from the sky, we would eat poison in the form of fertilizers and breathe asbestos every day. If something does harm, it should be forbidden, but if it can also benefit us, we should at least control it.

But what does this mean for digitization?

Regulate the Economy!

The relationship between capitalism and freedom is paradoxical. The fact that we are free to decide how to spend our money is a fundamental prerequisite for the free market economy. But, at the same time, every company tries its best to reduce our choices, so we select solely their product. The large I.T. corporations have developed an extremely profitable business model from helping to reduce our options. That may sound relatively harmless for the economy, but it's of existential importance for the marketplace of opinions. Our response to this problem should not be to slow down progress itself. Stopping it in the West would be futile, anyway, since we cannot control the rest of the world. It would be self-damaging, because digitization and A.I. can produce a lot of good without exploiting personal data to manipulate and control us. It would be against our nature, which is geared toward progress, including technological progress. But it would also be unfair to those who still seek advancement within a state, or in global competition, because this is where technological progress offers many opportunities. Finally, it would be unwise with regard to China. After all, asserting oneself against this new power requires economic strength.

So, the challenge will be to set freedom-enhancing rules that retain the benefits of digitization and A.I. for the United States, which is home to the largest players in this field, that will also be easy to achieve.

But Europe can do it, too, thanks to the European Union's formative and economic power.

Disperse Power Concentration

The big tech companies have too much power. They know too much about too many people. They make too much money and have too many reserves. They spend too much on lobbying—more than any other industry—in Washington, in Brussels, watering down data protection laws and A.I. policies. And within these corporations, individuals have too much power. Mark Zuckerberg holds about 60 percent of the voting rights of Facebook. That's the same Zuckerberg who announced the end of privacy in 2010, but bought his house's neighboring properties to protect his own privacy. Larry Page and Sergey Brin together hold an equal share of Google's parent company Alphabet's voting rights. Although Jeff Bezos owns *only* 17 percent of Amazon's voting rights, he is—even after his divorce cost him a quarter of his shares—the wealthiest person in the world, as well as the owner of *The Washington Post*.

Even if one were not offended by the imminent silencing of humanity in and of itself, one would have to be alarmed by the dangers of Zuckerberg's and co.'s abuse of power. Extreme concentrations of power have existed in the past. They were already a danger then, but are even more worrisome today, when digital

power peculiarities mean that small changes in a platform's code have potentially enormous and immediate
effects on millions of people. If Facebook's algorithm
starts promoting hate messages, or if YouTube pushes
conspiracy videos, these will affect the perceptions of
billions of people, especially since both platforms already know who will be most susceptible.

We can break up these power structures. European or U.S. antitrust authorities should have prohibited Facebook's purchases of Instagram and WhatsApp. Today, the company ought to be broken up,
and both subsidiaries should be split. We need to
promote competition and prevent the concentration
of power by preventing the big players from simply
swallowing up every up-and-coming competitor. The
E.U. Commission has proven its ability to act in many
cases by imposing fines running into billions of euros
for breaches of its competition rules by the big tech
groups. The U.S. has also imposed hefty penalties. In
2019, Facebook was fined $5 billion for violating the
privacy of millions of users worldwide.

We should also force companies that benefit from
network effects to provide connectivity to their competitors. The network effect means that everyone
chooses the provider with the most users because that
is where the benefit is the greatest. That's why neither
MySpace nor the German variant StudiVZ, or later
Google+, stood a chance against Facebook. Interfaces
could reduce this effect by enabling users of a data-
sensitive service, such as Threema, to connect with

WhatsApp users. The mobile phone market has long been subject to such rules of interoperability. Any T-Mobile subscriber can call any AT&T mobile phone. Without such a requirement, AT&T would have become a monopoly.

A further step would be to treat certain services as common goods and subject them to corresponding obligations. Facebook, Amazon, and Google offer an infrastructure comparable to retail spaces or the road network. We cannot avoid them, and many economic livelihoods depend on them. It stands to reason that access to the leading digital marketplaces for products (Amazon) and opinions (Facebook), as well as the road signs leading to them (Google), must not be subject to the capriciousness of private companies. Like the state, these companies must be bound by the principle of equal access, which means that Facebook and Amazon must have a valid reason for excluding anyone from their platforms beyond their terms and conditions.

And beyond access, we must also regulate how these digital platforms operate.

But the European General Directive on Data Protection (GDPR) and similar laws in other countries have their limits. They do not solve the issue that I.T. companies are permitted to penetrate our innermost selves and earn money. That they can manipulate us in a way that we neither can comprehend nor foresee. A brilliant idea by U.S. professors Jack M. Balkin and Jonathan Zittrain goes one step further: they want to turn I.T. companies into data trustees. The companies would only be allowed to use data in our interest, not in their own. Like doctors' offices or law firms, they would have to keep the data confidential. They would not be allowed to use it for advertising or other forms of influence. The exploitation of personality profiles or our emotional state would be a thing of the past unless it would benefit us. This would mean that services such as Facebook and Google Maps would have to earn money differently. They might even end up costing something.

But we are already paying for them, and have been for a long time—with our data.

Prohibit Automated Recommendations

Facebook's and YouTube's recommendation algorithms facilitate extreme right-wing opinions and conspiracy theories and hatred of LGBTQIA+ people and women. Contributions from radical opponents of vaccination ("anti-vaxxers") are particularly dangerous—they lament a sinister alliance between the pharmaceutical industry and the medical profession, spreading the misconception that vaccination leads to autism, and contribute to the resurgence of diseases we thought eradicated, such as polio. Because of such videos and posts, many in Brazil believe that vaccinations will infect children with the dangerous Zika virus. Refusal to vaccinate puts not only one's own child at risk, but also those for whom vaccines do not work. In 2019, the World Health Organization declared people who oppose vaccination a global threat.

Algorithms from Facebook and YouTube must not be allowed to contribute to this development. Automated recommendations should be prohibited as long as they systematically promote extremist contributions that seek to retain people on their channels through highly emotional messages. Such a ban would also affect other, often harmless, or valuable recommendations, but such recommendations would not have to be generated automatically. They could just as easily be made by employees or other trusted parties. Or the creators of contributions might provide the platforms with tags or keywords that group together similar posts and offerings. A ban on recommendation algorithms would not be censorship, because all videos and posts would still be uploaded and viewed. Still, no one has a right to benefit from automated recommendations. Continuing to rely on self-regulation would be negligent. The past has shown that platforms, if left to their own devices, react too late—if at all. Companies simply have no incentive to restrict their recommendation algorithms since they earn money by keeping users on their platforms.

But I.T. companies should not be able to make money with such recommendation algorithms or with our data.

Turn I.T. Corporations into Data Trustees

There has been a lot of criticism of the European Union's General Data Protection Regulation (GDPR), yet it is one of the most successful laws that the European Union has ever enacted. It establishes a right to be forgotten, i.e. the possibility of having all personal data held by a company deleted. It extends the rights to data disclosure, and it guarantees a right to take one's data to a competitor. Most important, however, are the regulations that the GDPR places on damages and fines for data-protection violations, not so much because it hurts companies to have to pay a fine of up to 4 percent of their total global annual turnover, but because it underscores the value of our data in terms that even commercial enterprises understand. The success of the GDPR is also demonstrated by its global impact. The European Union is an important market, comparable in size to the U.S. or China, and its regulations cannot be ignored. All companies that offer products or services in the E.U. that require the collecting or processing of data must comply with the GDPR. Since the Cambridge Analytica scandal broke, even U.S. states, such as California, are using the GDPR as a model for their own data-protection legislation. If a federal data-protection law were ever passed in the U.S., it would have enormous positive impacts, reaching far beyond the American borders.

Control Algorithms and Anonymize Data

The business community likes to complain that European data-protection regulations put them at a competitive disadvantage against U.S. companies. This is nonsense. Europe would not have founded a European Apple, Facebook, Google, Microsoft, or Amazon, even without data protection. European countries have a different risk culture. They don't spend money as freely, and the European market is not as homogeneous as the U.S. market. European markets don't attract as many top academics, and the U.S. has the technological edge. Accordingly, the development of A.I. in Europe is not being hindered by regulations, but by other factors.

Algorithms must be monitored. In Germany, for example, all products are certified by the Technical Control Association. In the U.S., the Federal Communications Commission certifies electronic products. Algorithms should also have to comply with specific standards and be subject to adequate supervision, especially if they fulfill sensitive tasks. Sensitive applications might include use in the healthcare sector or by the police, but also the mass influencing of public opinion on Facebook, Twitter, or YouTube. Software developers must be held liable, regardless of whether they committed a mistake themselves. Such strict liability has long existed with, for example, motor vehicles or pet ownership.

Additionally, sensitive data used for the training

of A.I. should only be accessible in an anonymous form. This is not as easy as it sounds. Hackers have demonstrated how simple it is to link data to a specific person, even if names, dates of birth, or social security numbers have been removed. Nevertheless, anonymization is possible without giving up the benefits of machine learning or leaving the field to China. For instance, medical research is already working on training A.I. with sensitive data in a black box that no one can access. There are also mathematical procedures that can modify data records to such an extent that they are suitable for machine learning, but otherwise unusable. Apple calls its use of such methods "differential privacy" and claims to have reconciled the possibilities of A.I. with respect for the privacy of its customers. Whether this is true or not is something no one seems to be checking at the moment. It is up to us to demand laws that enforce such regulations.

And while we're at it, we should also reduce the testosterone levels of the security state.

Limit State Power!

Surveillance is only a symptom. The disease is the fueling of irrational fears. Scaring people—whether about crime, extremist violence, or alienation—costs nothing. At the same time, the number of crimes recorded in my home country of Germany, for example, is lower than it has been since 1992. My country is extremely unlikely to become the victim of a large-scale terrorist attack, and foreigners are not evil, just different. It costs politicians nothing to "alleviate" the fears they have stoked through stringent security laws and stricter asylum laws, but every tightening of the law robs us of a little bit of freedom, takes away some of our humanity, and shifts more control from the individual to the state.

Digitization favors the concentration of power not only in the economy, but also in politics. Suddenly, a few can carry out and supervise many tasks simultaneously, including automated weapons systems and the use of force. This great power in the hands of a few must be constrained by strict limits on the use of digital technology. China shows us how effective digital technology can be when used to establish the perfect dictatorship.

Our safety becomes completely dehumanized if we leave it to A.I., instead of being responsible for it ourselves. Right now, A.I. holds enormous dangers, which we must identify and contain before it's deployed.

We Must Test, Monitor, and Control All A.I.

Reality often falls short of our aspirations. By the fall of 2019, six people had died in autonomous-vehicle tests in the U.S. In some cases, this was caused by autopilot failure. Self-driving cars will nevertheless end up causing fewer deaths than those driven by humans. They will—once they are widely used—undoubtedly make traffic safer. But it is important to remember the fallibility of artificial intelligence. The rampant use of A.I. by the judiciary and police is premature because artificial intelligence still shows itself to be too artificial and less than intelligent in far too many cases.

The data used to train A.I. to recognize patterns is, itself, full of prejudices. These prejudices will be digitized and baked into the system. Intelligent video cameras are up to ten times more likely to misidentify Black people's faces, especially Black women, than the faces of their white counterparts. Sheer statistics alone will lead to unacceptable numbers of false positives in automated searches for terrorists. If a computerized search for known terrorists were carried out among all 30 million people who use public transportation in Germany every day, even a very low error rate of 0.1 percent would lead to 30,000 false positives every day. Explain that to the person who got pulled out of their commute following a command from an officer's data goggles, and who has to establish their innocence by being questioned, searched, and fingerprinted.

A.I. should only fight crime if it knows how not to discriminate and if its failure rate is so low that it does more good than harm. It is not enough to be only as bad as its human counterparts. Because machines make decisions on a massive scale, they affect many more people, threatening to entrench shortcomings in the decision-making process because we question them less often.

The algorithm must never become sacred, and criticism of it must never be considered blasphemous. Those affected must always be able to question the results. To this end, algorithms must be transparent to those affected. Until algorithms function without error, every critical decision must be reviewed by a human—whether in applications of social welfare, court orders, or surveillance measures. To incentivize control, that person must be accountable, even under the threat of disciplinary action, for the results of the activity they monitor.

And A.I. and the State must never follow us everywhere simply to investigate us.

AfD Test, Index, and Expiration Date for New Safety Laws

The State is a strange creature. It serves and protects us, provides for us, and advances us, but at the same time, it's capable of a life of its own that can run counter to our interests. When things become personal— when a politician wants to force a political victory or cover up misconduct, when greed and lust for power take precedence over duty and decency—the common good suffers. Therefore, we cannot blindly trust public authorities, but must instead keep a very close eye on them.

This applies above all to the secret services and police. Their respect for our rights decides the fate of our democracy. Not only should police officers have to prove this respect in their everyday work, but we also need to see that respect reflected in the laws these officers are tasked with enforcing. The more a law permits, and the vaguer the criteria for suspending our fundamental rights, the greater the danger posed to our freedom. Such criteria increasingly leave the police and their leadership—not the courts—to decide whether or not they take action. Laws must, therefore, always minimize the risk of such abuse.

Every new authorization for police and secret service monitoring must pass the extremist test. Would we want to leave such authority in the hands of a future populist and right-wing government, such as by the German AfD? Vague terms such as the "imminent danger" contained in the Bavarian Police Act, or low intervention thresholds for intensive measures, such as secret online searches of smartphones and computers, all fail.

But even beyond the extremist test, we need limits on restrictions on freedom. The German Federal Constitutional Court once rightly demanded that new security laws be understood in the context of the total body of "surveillance law." It's not just the final step that needs to be assessed, but the whole path leading to that final step. Such an approach is, however, extremely complicated. How do we determine the importance of each item of data? How does su-

perficial surveillance of the whole weigh against the
close monitoring of a few? Does it depend on legal
authority or practical use? And when the data's value
is determined, how do we decide how much is too
much? In legal practice, the approach to consider the
sum of surveillance authorizations has not yet played
any role. This must change, despite the difficulties
mentioned above. We should index surveillance powers. We could set a value of 100 for the sum of authorizations at a certain point in time—and set a maximum value for the future, for example, 102. New
laws would then have to be assessed by independent
experts against all existing levels of monitoring. If a
new authorization exceeds the maximum value, the
legislature would first have to abolish an old monitoring measure before introducing the new one.

We also need a default expiration date for security
laws and to avoid retaining restrictions that have out-
lived their practical use. This should be self-evident.
The rule of law is based on the principle that every
encroachment on fundamental rights must be justi-
fied. In practice, however, the German legislature
avoids this principle far too often by relying on its
so-called prerogative of assessment. Because of this
leeway granted by the Federal Constitutional Court,
parliament can enact laws of uncertain effectiveness.
One such example is the mass evaluation of passenger data, as mentioned above. Yet once the efficacy
can be verified, this leeway should no longer apply.
After two years at most, every security law should

be subjected to an expert examination and be imme-
diately repealed if its effectiveness cannot be proven.

This would also reopen virtual, as well as real,
spaces the state has unjustly closed.

Preserving Open Spaces

A democracy needs space for discussion, information,
and the free expression of the individual, as well as for
behavior and opinions that the majority condemns,
without them being punishable. People must be able
to seek shelter when they are politically persecuted.
People must be able to confide in the press without
fear of retribution and blow the whistle on abuses
in companies or within government agencies. They
must be able to maintain contact with politically like-
minded people without discrimination. They must be
able to inform themselves about diseases and how to
deal with addiction, the burdens of raising children,
and frustration at work, without being pilloried or
otherwise punished.

The state must maintain these essential freedoms
instead of limiting them. This is especially true in
virtual space, where minorities, journalists, whistle-
blowers, and those seeking advice can anonymously
exercise their rights. In the context of these fundamental rights, politicians' demands to allow only the
use of one's own clear name on the internet are very
dangerous.

Anonymity is protection. Penalties for the operation of so-called Tor networks, which allow anonymous surfing on the internet, but which also harbor the "darknet," in which drugs and weapons and child pornography can be found, are too shortsighted. The vast majority of Tor networks have legitimate reasons for their activities, including to escape corporate surveillance or internet censorship by autocratic governments. We don't close all public parks just because drugs are sold in some of them, so the rules for these virtual spaces should be the same.

What we should seek to close are the deep trenches that run through our society.

Counter-Extremism by Design

We cannot rely on the courts to protect our freedom. Courts annul laws only if they go beyond the boundaries of the constitution. But our challenge to politics and to ourselves must not be to push freedom to the limits—and thus, necessarily, beyond them. We would satisfy a higher standard by implementing the above proposals, and by working toward a political system that makes the path to implementation for extremist positions as awkward as possible. A prime guarantor for this is proportional representation.

In a democracy with proportional representation, a parliament reflects the approximate percentages of votes different parties have won. Germany has one such electoral system. In the opposite model, majority rule (or "first past the post" voting), only one party wins at a time, namely the party with the most votes in a constituency. All other votes are irrelevant. The U.S. is an example of this type of electoral system. The majority voting system favors a two-party system, while proportional representation tends toward a multi-party system.

There are many arguments for and against one electoral system or another. In the age of digitization, however, the most severe effect of a two-party system is that it poisons the discourse. Developments in the U.S. illustrate this. All major issues have been reduced to an either/or proposition: man-made climate change either exists or doesn't exist. Guns kill or they're used for self-defense. Abortion is an expression of freedom or murder. Immigration is an opportunity or a grave threat. Everything is reduced to victory or defeat, right or wrong, yes or no, friend or foe. The system leaves no room for nuance, and there's no room for a broad spectrum of solutions to the increasingly complex challenges of our times. It is as binary as a computer that thinks in zeros and ones.

Now it is true that there are extreme opinions even in countries with proportional representation, and polarization occurs in these climates, too. However, these opinions remain clearly recognizable as extreme positions and are reserved for extremist parties' platforms. There is less danger of centrist parties adopting these positions, as the Republican Party in the U.S. did after Donald Trump's election in 2016. It's very difficult for extremist ideas to achieve an absolute majority in a parliamentary system. Even Adolf Hitler's party never gained an absolute majority, until he formed a one-party state.

Therefore, states that do not yet have a proportional representation system should introduce it, and all others should safeguard it for the future. The German constitution does not guarantee proportional representation. That must change.

These and similar proposals are not guaranteed to prevent extremist policies. However, obstacles to achieving an absolute majority are essential because they slow down the State's takeover so that protests and a renewed power shift remain possible before it's too late.

But to do this, we must hurry.

Time is Running Out

A statue of George Orwell sitting in front of the BBC headquarters in London has an inscription that reads, "If freedom means anything at all, it is the right to tell others what they do not want to hear."

That's what this volume does. That's what I do in my daily work as an attorney for the German Society for Civil Rights, which takes constitutional violations to court, but also as a citizen who demands that we must fight for our freedom.

This fight is not against robots. The great danger of artificial intelligence is not that it could one day subjugate humanity, but instead that it could be used by us humans to destroy freedom. I am therefore not arguing against a transfer of responsibility to machines. I plead against a transfer of responsibility to machines that are controlled by humans whose interests do not align with the public's interests; to machines that discriminate and destroy our sense of self, our self-confidence; to machines that totally capture us, predict our behavior, put us in pigeonholes, and keep us there.

We can prevent all this from happening. We still know life in freedom, but our descendants could be born into a zoo, ignorant and comfortable. They would know nothing but a life in which invisible higher intelligence permanently observes, cares for, entertains, and protects them from their environment's adversities.

If we don't want such a life for our children, if we believe that they, and all people, are happier in freedom than under the supervision of algorithms—if we are to maintain their dignity—then we have to act!

Yes, we are all capable of being manipulated, are easily frightened, and lazy. But we are also capable of recognizing this tendency and capable of rebelling against being controlled by companies, by right-wing populism, or by the State. We have to question whether the offerings of I.T. companies and security politicians really do us more good than harm. We must formulate demands like those I've outlined here, and choose and support candidates and parties that protect freedom. We must use our market power and our western ideas and tools to defend our values against China, and against countries that seek to emulate them. We must do without data-hungry "free" services, and be willing to pay for services that protect our privacy and freedom.

Every one of us has to step up to save mankind's greatest achievement: our self-determination, our freedom!

BIJAN MOINI, born near Karlsruhe, Germany, is a lawyer, political scientist, and civil rights campaigner with German-Iranian roots. After studying and training in Hong Kong and Berlin, he worked as a lawyer for a commercial law firm. He now coordinates constitutional complaints for a Berlin civil rights NGO and writes and speaks on sociopolitical subjects such as digitization, surveillance, and data protection. He is the author of the novel *The Cube* about a near-future in which society has traded access to its most personal information for an ideal world—until it all goes wrong. Moini lives in Berlin, Germany with his family.